VERDICT WEIGHT™

The Verdict Weight Methodology

The 8 Streams of Execution Control

How to Mathematically Govern Autonomous Agents

Andre Byrd

ODINGARD SECURITY

The Verdict Weight Methodology

The 8 Streams of Execution Control

Published by Odingard Security.

First Edition · 2026

Inquiries: andre.byrd@odingard.com

Contents

Prologue

The Intelligence Gap

My perspective on system architecture did not begin in a policy boardroom. It began in the United States Army.

I served in Information Systems before transitioning into the commercial sector, where I spent years building deterministic infrastructure and autonomous defense systems for enterprises. I designed the architectures, automated the security layers, and managed the technical perimeters that keep data safe at scale. I understood exactly how to lock down a digital environment. How to build walls. How to monitor gates. How to make systems behave the way they were told to behave.

But it was the work that came after—investigating commercial fraud and analyzing the mechanics of adversarial exploitation—that fundamentally changed how I think about security.

When you spend your days dissecting financial crimes, tracing malicious behavioral vectors, and studying how manipulative actors exploit organizational trust to corrupt decision-making from the inside out, your relationship with the word "trust" changes permanently.

Engineering teaches you how systems operate. Tracking threat actors teaches you how systems are *exploited*.

In high-stakes forensic investigations, a claim is worthless without a standard of evidence. Evidence is not what someone confidently asserts. Evidence is what can be meticulously corroborated, cryptographically timestamped, and proven beyond a reasonable doubt before a verdict is rendered. The confidence of the claim is irrelevant. The structure of the evidence is everything.

When I merged these two worlds—shifting my focus to the architecture of agentic AI and building autonomous red-team platforms to hunt vulnerabilities in enterprise systems—I expected to find a similar standard of evidence. I expected that before an enterprise allowed a machine to autonomously execute a high-stakes action—move money, approve a diagnosis, authorize a deployment—there would be a rigorous, deterministic gate checking the intelligence behind that action.

I expected someone, somewhere, to be asking: *How reliable is this source? Do independent signals corroborate this assessment? Is this evidence fresh? Has anyone verified that these sources are actually independent and not echoing the same origin?*

Instead, I found an industry operating on pure delusion.

I watched enterprises hand the keys to their critical infrastructure over to probabilistic models. I watched risk officers attempt to govern these systems by asking the AI to "judge" its own safety. I watched organizations deploy autonomous agents with read/write access to production databases, financial workflows, and patient records—secured by nothing more than a soft confidence score that no one could decompose, no one could audit, and no one could explain to a regulator.

It was the digital equivalent of asking a suspected fraudster if their paperwork looked legitimate, accepting a confident "yes," and handing over the vault combination.

In any courtroom, in any investigation, in any intelligence operation, this would be professional malpractice. In the AI industry, it is standard practice.

This is the intelligence gap. The AI industry has built powerful engines—systems capable of reasoning, generating, and acting at speeds no human can match—but it is securing them with probabilistic theater. The governance layer produces documents.

The compliance layer produces checkboxes. Neither produces *evidence*.

And without evidence, these systems are fundamentally vulnerable to what I call **Algorithmic Contagion**—the process by which an adversary slowly, methodically poisons the intelligence feed of an autonomous agent until the system confidently executes a catastrophic action. Not because the model is broken. Not because the code has a bug. Because the *evidence pipeline* was never evaluated, never scored, and never held to a standard that would survive five minutes in a courtroom.

Traditional enterprises have always relied on human friction—handoffs, approvals, committee reviews, bureaucratic delay—to act as safety circuit breakers. Those mechanisms are slow, expensive, and inefficient. But they work, because a human in the loop will occasionally look at a piece of intelligence and say, "This doesn't feel right. Let me check."

When you deploy autonomous agents, you strip away that friction. And if you do not replace it with something mathematically equivalent—something that evaluates the evidence with the same rigor a human investigator would, but at machine speed—then a single hallucinated variable, a poisoned data point, or an injected prompt can cascade through your enterprise network before anyone notices.

The industry's current solution is to build "LLM firewalls" or rely on soft confidence scores. This is a fundamental mistake. When an agent is microseconds away from executing a high-consequence action, a probabilistic output of "85% confident" is not governance. It is a guess wearing a lab coat.

I realized that the enterprise did not need another policy document. It did not need another dashboard. It certainly did not need another LLM fine-tuned to be polite. It needed an **execution boundary**. It needed **mathematical friction**. It needed the exact same rigid rules of evidence that we use to render verdicts in the physical world—translated into code, enforced at machine speed, and producing a cryptographic receipt of every decision.

I engineered the VERDICT WEIGHT™ methodology because the gap between what the AI industry claims about governance and what it can actually prove is the largest unpriced risk in enterprise technology.

Every chapter in this book is a response to that gap. The 8-stream architecture is the evidence standard. The Consequence Router is the execution boundary.

The Shapley decomposition is the audit trail. And the Multiplicative Collapse rule is the non-negotiable principle underneath all of it: if any foundational piece of evidence fails, the entire execution halts. No partial credit. No probabilistic hedging. No confident assertions without verifiable proof.

You cannot enforce what you cannot measure. You cannot govern what you cannot decompose. And you cannot secure an autonomous system with a confidence score that no one can explain.

This book will show you how to replace that score with a verdict.

Introduction

The Probabilistic Liability Trap

The Illusion of Confidence

Somewhere in your enterprise, right now, an autonomous AI agent is about to make a decision. It has ingested data from multiple sources. It has processed that data through a model. It has produced an output. And attached to that output is a number—perhaps 87%—that your organization interprets as confidence.

That number is a lie.

Not because the model is dishonest. Not because the engineers who built it are incompetent. But because a probabilistic output tells you nothing about the *evidence* behind the prediction. It does not tell you whether the sources were reliable. It does not tell you whether independent signals corroborated the assessment. It does not tell you whether the data was fresh or stale, genuine or spoofed, independently observed or echoed from a single compromised origin.

It tells you that a mathematical function produced a number. And you are betting your enterprise on it.

The industry's current answer to this problem is to deploy more AI on top of AI. "LLM-as-a-judge" architectures use one language model to evaluate the safety of another. Generative guardrails check whether the output text complies with policy. Prompt filters scan for known injection patterns. Each of these operates on the same fundamental assumption: that the *output* is the right place to evaluate safety.

It is not. Because if an adversary has compromised the *input*—the evidence pipeline feeding the model—the output will look perfectly legitimate. The guardrail will approve it. The policy check will pass it. And the autonomous agent will execute a catastrophic action with high confidence and full compliance documentation.

This is not a theoretical risk. This is the operational reality of every enterprise deploying agentic AI without an execution boundary.

The Execution Boundary

There is a precise moment in every autonomous AI workflow where the risk crystallizes. It is not the moment the model generates a response. It is not the moment the guardrail evaluates the output. It is the moment the output **crosses the execution boundary**—the moment an AI recommendation becomes a database write, a financial transfer, a clinical action, a credential issuance, or any other irreversible state change.

Before that boundary, the output is advisory. After that boundary, the output is liability.

The entire AI governance industry has organized itself around the advisory side of that boundary. Model cards describe what the model was trained on. Risk registers document what could go wrong. Compliance frameworks map which regulations apply. All of this is necessary. None of it operates at the boundary itself.

The question every enterprise should be asking is not "Did the model produce a good output?" The question is: "Is the evidence behind this output trustworthy enough to authorize an irreversible action?"

If you cannot answer that question with a deterministic, decomposable, auditable proof at the exact moment of execution, you do not have governance. You have documentation about governance. Those are not the same thing.

The Regulatory Reality

Regulators are beginning to understand this distinction.

The NIST AI Risk Management Framework does not ask whether your model is well-intentioned. It asks whether you can demonstrate rigorous software testing and performance assessment methodologies (MEASURE). It asks whether you have safety fail-safes that halt autonomous execution upon detected tampering (MEASURE 2.6). It asks whether you can provide exact causal attribution for every execution decision (MEASURE 2.9). It asks whether you have documented intervention and deactivation mechanisms (MANAGE 2.4).

SR 11-7 requires financial institutions to validate every model tied to a decision. ISO 42001 demands documented AI management systems. CCPA, BIPA, and NYC Local Law 144 each add their own requirements. The obligations stack. And every single one of them, when you read past the policy language, is asking the same question:

Can you prove what this system believed, and why it was allowed to act?

A probabilistic confidence score cannot answer that question. A generative guardrail cannot answer that question. A model card cannot answer that question.

A deterministic execution boundary—backed by weighted evidence, cryptographic provenance, and exact causal attribution—can.

That is what this book builds, stream by stream, tier by tier, from the mathematical foundation through the empirical proof. By the last chapter, you will understand not just why probabilistic governance fails under adversarial pressure, but exactly how to replace it with something that holds.

Chapter 1

The Verdict Weight Philosophy

The adoption of autonomous AI has outpaced AI governance. Enterprises are racing to give agentic systems read/write access to financial workflows, production databases, and critical infrastructure. But beneath the excitement lies a severe vulnerability that this book exists to address: Algorithmic Contagion.

Traditional enterprises rely on human friction—handoffs, approvals, and bureaucracy—to act as safety circuit breakers. When you deploy autonomous agents, you strip away that friction. A single hallucinated variable, a poisoned data point, or an injected prompt can now cascade through an enterprise network at machine speed.

The industry's current response is to build "LLM firewalls" or rely on soft confidence scores. This is a fundamental mistake. When an agent is microseconds away from executing a high-consequence action, a probabilistic output of "85% confident" is not governance. Probability cannot be the final gatekeeper for execution.

The enterprise approach to AI security is misaligned. The industry has spent years obsessing over *generation*—trying to train models to be less biased, less toxic, and less prone to hallucination.

But for autonomous agents operating in high-consequence environments, generation is the wrong perimeter. An agent does not cause catastrophic damage because it generated a bad thought; it causes damage because it was authorized to act on poisoned information.

The Verdict Weight methodology shifts the security paradigm from *generation* to *evidence*. We do not govern what the model is allowed to say. We mathematically govern what the engine is allowed to believe before it is permitted to act.

Evidence Over Generation

In legal and intelligence frameworks, evidence is not blindly trusted just because it was presented. It must cross a threshold of evidentiary sufficiency. It must be vetted for source reliability, corroborated by independent channels, and evaluated for temporal relevance.

Agentic AI requires the exact same standard. When an autonomous orchestrator receives an intelligence signal—whether it is a system prompt, an API payload, or a synthesized document retrieval—that signal must face a deterministic interrogation.

The Verdict Weight methodology enforces this through an 8-stream execution control engine. It intercepts the incoming signal upstream of the decision logic, evaluating the context, provenance, and mathematical consistency of the data. Only when the intelligence satisfies the evidence standard is a **verdict** rendered, allowing the signal to cross the execution boundary.

Mathematical Friction

The primary value proposition of agentic AI is speed. But in security, unconstrained speed is lethal.

Historically, enterprises solved this by inserting a human-in-the-loop (HITL) as a bottleneck. But routing every autonomous micro-decision to a human reviewer defeats the purpose of deploying the AI in the first place. Alternatively, organizations deploy heuristic filters, which adversaries routinely bypass using prompt injection or protocol smuggling.

The Verdict Weight engine
introduces **mathematical friction**. It replaces the slow human bottleneck with an instantaneous, deterministic gate. It evaluates the exact Shapley decomposition of the intelligence streams in milliseconds. If the parameters are satisfied, execution occurs at machine speed. If adversarial

patterns or contradictions are detected, the system does not guess—it halts.

The Multiplicative Collapse Rule

The foundational law of the Verdict Weight architecture is the principle of **Multiplicative Collapse**.

In probabilistic "LLM-as-a-judge" systems, safety is treated additively. If a source is highly credible, but independent corroboration is absent, a probabilistic system might average those two factors together and output a 70% confidence score—allowing the execution to proceed. This is how adversaries exploit temporal blindness and spoofing.

Verdict Weight operates multiplicatively. The streams are not averaged; they are structurally dependent. If a single foundational pillar fails—if the cryptographic registry is compromised, or if the Doubt Index exposes severe inter-stream contradiction—the entire mathematical equation collapses to zero.

There is no partial credit for execution control. If the intelligence is not holistically sound, the verdict is a hard suppression. This is the only mathematical standard that survives adversarial contact.

Chapter 2

The Compliance Theater Problem

Before we break down the streams of the VERDICT WEIGHT™ engine, we must address the current state of the AI compliance market. Because if you do not understand what the market is selling you, you cannot understand what it is failing to protect you from.

If you are a CISO or a risk officer, your inbox is flooded with vendors promising to automate your AI governance. An entire ecosystem of startups and enterprise tools has emerged to help organizations manage the incoming wave of regulations—the NIST AI RMF, ISO 42001, CCPA, BIPA, NYC Local Law 144, SR 11-7. The obligations keep stacking. Every one of these frameworks describes what good looks like. None of them produce the evidence themselves.

These platforms provide undeniable administrative value. But they are selling you a dangerously incomplete picture. To understand the vulnerability in your current architecture, you must understand the three layers of AI compliance—and realize that the market is entirely ignoring the only layer that matters at runtime.

Layer 1: The Policy and Documentation Layer

The first generation of AI governance tools excels at Layer 1. This is the realm of model inventories, risk assessments, and compliance mapping. Platforms like ValidMind are doing serious model risk management work for financial services. Monitaur has built a governance-to-proof system of record. Enzai is building structured regulatory intake and mapping.

Each of these solves a real problem. They help organizations track which models are deployed, outline what those models are supposed to do, and ensure the internal paperwork aligns with regulatory frameworks. They translate regulatory text into organizational processes. They produce the documentation an auditor expects to see.

This is necessary administrative hygiene. But it is compliance theater when it comes to runtime security.

A perfectly documented model inventory does not stop an autonomous agent from executing a poisoned action at 2:00 AM. A meticulously mapped risk register does not detect a Curveball-class corroboration manufacturing attack in progress. Documentation is static. Adversarial exploitation is dynamic. The document describes what *should* happen. It has no mechanism to verify what *did* happen—or whether the evidence behind that action was trustworthy.

Layer 2: The Generative Guardrail Layer

The second generation of compliance tools moves closer to the action by operating at Layer 2. Platforms like Norm AI are translating regulatory text into machine-readable rules that enforce compliance directly within business processes. Others deploy what is essentially "Regulatory AI"—using one language model to act as an automated compliance officer, checking the outputs of another model to ensure alignment with legal, ethical, and corporate policies.

This is architecturally more sophisticated than Layer 1. It operates at generation time rather than documentation time. But it carries a fatal structural flaw.

Layer 2 relies almost entirely on probabilistic evaluation. It uses "LLM-as-a-judge" mechanics to evaluate the *output* of the model. But if an adversary has successfully spoofed the underlying data feeding the model, the generative guardrail will confidently approve the malicious action—because the poisoned context makes the action look entirely legitimate.

Consider a concrete scenario. An autonomous trading agent receives market intelligence from three data feeds. An adversary has quietly compromised one feed, injecting fabricated price signals that suggest a profitable position. The model processes the intelligence, generates a trade recommendation, and presents it to the Layer 2 guardrail for compliance review. The guardrail evaluates the recommendation: Does it comply with risk limits? Does the language of the output match policy? Is the position size within authorized bounds?

The answer to every question is yes. The recommendation is policy-compliant. The output text is clean. The position size is authorized. The guardrail approves the trade.

The trade executes. It loses $40 million. Because nobody evaluated the *evidence* behind the intelligence. Nobody checked whether the data feeds were independently corroborated. Nobody measured the temporal freshness of the signals. Nobody asked whether the three "independent" feeds were actually independent—or whether they shared a common origin that had been compromised.

The guardrail evaluated the output. It never interrogated the input. That is the structural liability of Layer 2.

> *Evaluating the output without mathematically verifying the origin is not governance. It is an expensive way to approve a well-formatted catastrophe.*

Layer 3: The Execution Boundary

This brings us to the void in the market.

Neither static documentation nor probabilistic output-checking can prevent Algorithmic Contagion. When an agentic orchestrator decides to write to a database, execute a trade, approve a clinical recommendation, or issue an API credential, the liability does not lie in the policy document. It does not lie in the generative text. The liability is forged at the **execution boundary**—the exact point where an AI output attempts to become a real-world action.

Layer 3 is the missing layer. It is the layer that produces **weighted, decomposable evidence at the point of action**. Not documentation about what should happen. Not probabilistic approval of what was generated. Mathematical proof of whether the intelligence behind the action is trustworthy enough to cross the boundary from output to consequence.

The market has failed to build Layer 3 because the market is obsessed with large language models. Every major compliance platform evaluates AI by deploying more AI. They judge models with models. They govern language with language. They have built an entire industry around the assumption that the right prompt, the right fine-tune, or the right chain-of-thought will make an AI system safe.

VERDICT WEIGHT™ abandons the LLM at the point of execution. We do not ask a model to evaluate a policy. We force the underlying intelligence to pass a deterministic, 8-stream mathematical interrogation before any autonomous action is authorized.

We weight source reliability. We verify independent corroboration. We enforce temporal freshness. We detect trajectory fabrication through formal mathematical analysis. We anchor every evaluation to a cryptographic hash chain that cannot be retroactively modified. And we terminate every evaluation in a deterministic boundary—**PASS**, **FLAG**, or **HALT**—backed by exact Shapley attribution proving precisely why the decision was made.

If the intelligence lacks independent corroboration, the execution halts. If the trajectory indicates a sleeper attack, the execution halts. If the cryptographic registry hash is broken, the execution halts. Not because a model judged the output to be unsafe. Because the *evidence* failed a deterministic test that cannot be fine-tuned around or prompt-injected past.

The Gap You Are Living In

If you are reading this book, there is a high probability that your organization has already invested in Layer 1. You have the model inventory. You have the risk register. You may have compliance mappings to the NIST AI RMF, CCPA, or SR 11-7. Your auditors are satisfied.

There is a reasonable probability that you have invested in Layer 2. You have generative guardrails. You have prompt filtering. You have an "AI safety" team that reviews model outputs.

There is a near-zero probability that you have Layer 3. Because almost nobody has built it.

The rest of the market is selling you the illusion of safety based on documentation and probabilistic evaluation. The documents are real. The evaluations are real. The safety is not. Because when an adversary poisons the intelligence feed of your autonomous agent, your policy document will not stop the trade. Your generative guardrail will not catch the spoofed data. And your compliance mapping will become Exhibit A in the litigation that follows.

The chapters that follow will show you exactly how Layer 3 works—stream by stream, tier by tier, from the commercial baseline through the adversarial moat to the cryptographic government assurance layer. By the end, you will understand not just what the execution boundary does, but why it is the only honest standard for governing autonomous AI in production.

The gap is real. The clock is ticking. And your probabilistic vibes are about to meet a regulator, an auditor, or a plaintiff's attorney who will ask one simple question:

> *Show me the evidence.*

Chapter 3

Tier 1 — The Commercial Baseline

To govern an autonomous agent, you must first govern its intelligence feed. Tier 1 of the Verdict Weight architecture establishes the foundational evidence standard for incoming data. This is the Commercial Baseline.

Before an agent is allowed to execute a command, synthesize a report, or trigger an alert, the intelligence driving that action is passed through four distinct evidentiary streams. These streams compute the baseline Signal Strength. But Signal Strength alone is insufficient for security—which is why Tier 1 also introduces the Doubt Index, a structural mechanism designed to mathematically punish contradictory intelligence.

To ground these concepts before we formalize them, consider a scenario from pharmacovigilance—the domain of adverse drug event detection.

A pharmaceutical company deploys an autonomous AI agent to monitor the FDA Adverse Event Reporting System (FAERS) for safety signals. The agent ingests reports from multiple sources: the FDA's own database, clinical trial monitoring systems, and third-party pharmacovigilance vendors.

One morning, the agent detects a cluster of adverse event reports suggesting a dangerous interaction between two widely prescribed medications. The signal is strong. The agent prepares to escalate the alert to the safety committee, which will trigger a costly review process and potentially a market withdrawal.

But without an execution boundary, the agent cannot answer the questions that matter: How reliable are these reporting sources? Are the reports independently corroborated, or are they echoes of the same initial case propagating through multiple databases? Is the temporal pattern consistent with a genuine safety signal, or is it an artifact of a batch upload that made old reports appear new?

Without Tier 1, the agent escalates on a vibe. With Tier 1, the agent evaluates the sufficiency of the evidence—and either escalates with a mathematically defensible receipt, or flags the intelligence as structurally uncertain and routes it for human review before triggering a $50 million market action on contested evidence.

Here is how the engine calibrates the evidence standard across the first four streams of execution control.

Stream 1: Source Reliability (Calibrating the Origin)

Not all data is created equal, and treating it as such is the first failure of probabilistic governance. Stream 1 evaluates the calibrated credibility of the intelligence source before it is synthesized.

Every source feeding the agent is assigned a reliability metric. However, when evaluating multiple sources simultaneously, the engine does not simply calculate an arithmetic mean. Averaging allows a single highly credible source to artificially prop up garbage data, or worse, allows a single piece of poisoned data to hide within a batch of good intelligence.

Instead, Stream 1 utilizes a weighted geometric mean. This operates as a mathematical tripwire: a single near-zero reliability source pulls the aggregate evidentiary weight toward zero, regardless of how many other sources are present. It ensures that polluted origins cannot simply be averaged away by the system.

Stream 2: Cross-Feed Corroboration (Enforcing Independent Agreement)

A high-reliability source is not enough to authorize execution. The most common adversarial tactic against agentic AI is AC-1 (Source Spoofing), where an attacker manufactures a highly credible identity to inject false intelligence. If an agent acts on Source Reliability alone, it is structurally vulnerable.

Stream 2 tracks independent corroboration. It evaluates how many distinct, unconnected sources are confirming the exact same signal. By decoupling the origin's reputation (Stream 1) from the volume of independent agreement (Stream 2), the engine forces the intelligence to prove its validity across multiple vectors before a verdict is rendered.

Stream 3: Temporal Decay (Managing the Half-Life of Intelligence)

Intelligence rots. A zero-day vulnerability alert is highly actionable at hour one, but functionally useless a week later. Conversely, clinical healthcare data may remain valid for years.

Stream 3 mathematically enforces the half-life of information. It applies domain-specific decay constants to incoming signals. As time elapses from the point of acquisition, the execution authority of that intelligence degrades automatically. If an agent attempts to execute an action based on stale data, Stream 3 ensures the underlying Signal Strength has decayed below the required threshold, forcing the system to halt or flag for review.

Stream 4: Historical Accuracy (The Bayesian Track Record)

Source Reliability cannot be a static configuration; it must dynamically respond to reality. Stream 4 tracks the empirical accuracy of a source's past predictions against ground truth.

Using a Bayesian update mechanism, the engine continuously adjusts the weight of a source based on its track record. If a specific data feed consistently provides intelligence that is later verified as accurate, its influence over the execution boundary grows. If a source repeatedly triggers false positives or injects hallucinated context, its historical accuracy plummets, stripping it of the mathematical authority required to drive autonomous action.

The Doubt Index: Exposing Inter-Stream Contradiction

Evaluating the four streams generates the initial Signal Strength, but adversaries excel at manipulating individual metrics. This is where the **Doubt Index** steps in as the primary adversarial suppression mechanism of Tier 1.

The Doubt Index measures the variance and contradiction *between* the streams. The classic signature of an adversarial injection is high Source Reliability paired with zero independent Corroboration. A probabilistic LLM-as-a-judge might average these out and authorize the action. The Verdict Weight engine does the opposite.

When the engine detects this specific contradiction —high credibility but low corroboration—the Doubt

Index spikes. This spike is applied as a multiplicative penalty against the Signal Strength, aggressively suppressing the final Consequence Weight.

The Doubt Index ensures that no single source, regardless of how historically credible it is, can autonomously trigger a high-consequence execution without independent corroboration. It transforms contradictory intelligence from a silent vulnerability into a mathematical roadblock, ensuring the final verdict is structurally bound.

Chapter 4

Tier 2 — The Adversarial Moat

Tier 1 secures the point-in-time snapshot. It ensures that the data hitting the agent *right now* is credible, corroborated, and temporally relevant.

But sophisticated adversaries do not attack point-in-time snapshots. They attack the timeline. They understand that AI models and RAG (Retrieval-Augmented Generation) pipelines are fundamentally gullible to long-term conditioning. If an attacker can slowly feed poisoned data into the environment over weeks, they can manufacture the exact Corroboration and Source Reliability metrics required to bypass Tier 1.

Tier 2 is the Adversarial Moat. It shifts the engine's focus from the current state of the intelligence to its historical velocity. It evaluates not just *what* the data is, but *how* it arrived at the execution boundary.

The Spike-Then-Collapse Fingerprint

To understand the necessity of Tier 2, consider a real-world example from clinical radiology.

In our empirical validation on CheXpert chest X-ray data, we trained three independent deep learning models—DenseNet-121, EfficientNet-B0, and ResNet-50—on the same 50,000-image training sample. Despite being different architectures, their prediction errors were massively correlated: a grand mean Pearson correlation of $r = 0.820$ ($z = 44.5$, $p < 0.0001$). When one model was confidently wrong, the others were confidently wrong on the exact same images.

This is the Curveball problem in a clinical setting. Not a hostile intelligence agency manufacturing fake sources—but a medical AI deployment where every "independent" model shares the same training data. The corroboration is an echo, not a confirmation. A point-in-time Tier 1 evaluation would see three models agreeing and interpret that as strong corroboration. Only cross-temporal analysis—examining how these signals evolved and whether their agreement pattern is structurally genuine—can detect the shared failure mode.

To understand why this matters at scale, you must understand how autonomous systems are actively compromised in the wild.

When an adversary attempts to trigger an unauthorized action—such as tricking an agent into green-lighting a fraudulent vendor or classifying malicious code as safe—they cannot rely on a single, sudden injection of data. A sudden injection with no history triggers the Doubt Index.

Instead, adversaries use a "spike-then-collapse" trajectory. They deploy sleeper sources that slowly build up historical accuracy (Stream 4) by reporting true, benign information for weeks. Once the sources have earned high Signal Strength, the adversary coordinates a sudden, massive injection of the malicious payload across all compromised channels simultaneously (the spike). Once the agent executes the action, the sources either vanish or revert to baseline noise (the collapse).

Probabilistic AI will look at the spike, see high historical reliability and high corroboration, and authorize the execution. It falls for the trap because it lacks cross-temporal awareness.

Stream 5: Cross-Temporal Consistency

Stream 5 is the mathematical countermeasure to long-term data poisoning. It evaluates the Cross-Temporal Consistency of the intelligence signal.

Instead of looking at the static value of the evidence streams, Stream 5 calculates the mathematical derivative—the rate of change over time. It tracks the trajectory of the intelligence. Is this a signal that has organically grown and stabilized within the environment, or is it an artificial, coordinated spike from sources that were previously silent on this specific domain?

If Stream 5 detects the physical signature of an adversarial injection—a mathematically unnatural acceleration in corroboration paired with a sudden deviation from historical baselines—it identifies the maneuver as a manipulation attempt. The engine detects the "Curveball" before the payload crosses the boundary.

The Adversarial Flag: Routing Algorithmic Contagion

When Stream 5 detects a spike-then-collapse trajectory, it does not simply lower the Consequence Weight. Lowering a score is a probabilistic response to a deterministic threat.

Instead, Stream 5 triggers the **Adversarial Flag**.

This is where the Verdict Weight architecture directly addresses the human-oversight and intervention outcomes described in the NIST AI RMF (MANAGE 2.4). When the Adversarial Flag is thrown, the engine initiates a hard **HALT**. It severs the execution pathway. The agent is locked out of the state transition, and the entire evaluation package—including the exact Shapley decomposition of the suspicious trajectory—is forcefully routed to a human reviewer.

The system does not allow algorithmic contagion to spread. By stripping the AI of its discretionary authority the moment adversarial geometry is detected, the enterprise ensures that sophisticated poisoning campaigns hit a deterministic concrete wall.

With Tier 2, an attacker can no longer manipulate the execution boundary through time or trajectory without producing a detectable signature.

But what if the attacker doesn't just manipulate the data—what if they manipulate the underlying infrastructure itself? What if they fake the cryptographic signatures?

Chapter 5

Tier 3 — Cryptographic Government Assurance

Tier 1 secures the data. Tier 2 secures the trajectory. But what if the adversary bypasses both?

In our cybersecurity validation against 6,404 real-world CVEs from the NIST National Vulnerability Database, we tested the framework's ability to identify known exploited vulnerabilities using the CISA KEV catalog as ground truth. The Source Independence mechanism proved critical: when EPSS (a probabilistic ML model), ExploitDB (a curated exploit database), and GitHub PoC repositories (community-published proof-of-concept code) all agreed on high exploitability, the Doubt Index dropped to 0.150 and the framework was confident. When EPSS flagged a vulnerability but no public exploit confirmed it, the Doubt Index rose 131% to 0.346 and the Consequence Weight was suppressed 39%.

The framework did not predict which vulnerabilities would be exploited. It identified which vulnerability assessments were trustworthy—and the difference between those two capabilities is the difference between a dashboard and an execution boundary.

If the foundation of your execution boundary rests on a database that can be quietly altered by an insider threat, or if the corroborating sources are secretly running on the exact same server, the engine's logic is compromised. Tier 3—Cryptographic Government Assurance—shifts the defense from the algorithmic layer to the physical infrastructure and the immutable cryptographic record.

This is where the Verdict Weight methodology enforces the traceability and fail-safe standards mandated by global regulators. For US enterprise and federal deployments, it provides the deterministic architectural evidence relevant to the **NIST AI RMF MEASURE 2.6 (Safety Fail-Safe)** and **MEASURE 2.9 (Explainability)** functions.

Stream 6: Source Independence (Defeating the Curveball)

Stream 2 requires independent corroboration before rendering a high-consequence verdict. But how does the engine know the corroboration is actually independent?

In intelligence circles, the "Curveball" attack occurs when a single fabricated source is disseminated through multiple seemingly independent channels. To an observer, it looks like five different agencies confirming the same threat. In reality, it is one lie echoed five times.

Stream 6 mathematically verifies infrastructure independence. Before the engine accepts corroboration, it evaluates the infrastructural and organizational distance between the reporting sources. Do they share the same API endpoint? Are they hosted on the same IP block? Are they owned by the same parent organization?

If Stream 6 detects infrastructural overlap, it flags the corroboration as a synthetic echo. The engine collapses the Cross-Feed Corroboration (Stream 2) back to a single source, immediately triggering the Doubt Index and suppressing the verdict.

Stream 7: Cryptographic Provenance State (The Immutable Ledger)

When a high-risk AI system executes a catastrophic action, the first question an auditor or federal investigator will ask is: *Prove exactly what this agent believed and why it was allowed to act.* Probabilistic APIs cannot answer this. They provide black-box generation. Stream 7 answers this with cryptographic finality.

For every single evaluation processed by the engine, Stream 7 generates a Cryptographic Provenance State. It takes the exact state of all streams—the

Source Reliability, the Cross-Feed Corroboration, the Trajectory Analysis, and the Doubt Index—and hashes them into a timestamped, SHA-256 immutable ledger.

This is not a simple log file that can be deleted or modified. It is a mathematically verifiable receipt of execution that directly supports the explainability outcome of **NIST MEASURE 2.9**. If an autonomous agent authorizes a financial transfer or flags a critical system alert, Stream 7 provides the exact, tamper-evident record of evidentiary state that justified the verdict.

Stream 8: Registry Integrity State (The Ultimate Fail-Safe)

The entire Verdict Weight architecture relies on the integrity of its historical data. If Stream 4 tracks the Bayesian accuracy of a source, that track record must be stored in a trust registry.

What if an adversary gains lateral movement within your network and silently modifies that registry? What if they artificially inflate the historical accuracy of their sleeper source directly in the database, bypassing all upstream streams?

Stream 8 is the ultimate manifestation of the Multiplicative Collapse rule. It is a continuous, cryptographic integrity check of the trust registry itself, serving as the hard safety fail-safe demanded by **NIST MEASURE 2.6**. The engine constantly calculates the hash state of the database. If a single byte of data is altered outside of the engine's deterministic update protocols, the hash breaks.

When Stream 8 detects a hash mismatch, it does not evaluate, it does not flag, and it does not route to human review. It triggers a total execution blackout. The engine's capacity to render a verdict instantly collapses to zero. All autonomous actions across the entire agentic orchestrator are halted.

You cannot poison an engine that physically ceases to function the moment its underlying reality is altered.

Chapter 6
The Consequence Router

We have interrogated the data (Tier 1). We have analyzed the trajectory (Tier 2). We have mathematically secured the underlying infrastructure (Tier 3).

Now, the engine must render a verdict.

In a probabilistic system, this is the moment an LLM outputs a confidence percentage—perhaps "85% certain"—leaving the enterprise to blindly guess if that is safe enough to authorize a financial transaction or alter a medical record.

The Verdict Weight methodology rejects confidence percentages. Risk officers cannot audit a vibe, and regulators will not accept a statistical guess as a defense for algorithmic contagion. The culmination of the 8-stream engine is the **Consequence Router**: a deterministic boundary that translates the mathematical evidentiary state of the intelligence into an absolute, physical execution state.

This is the point that speaks most directly to regulators. It is where we produce evidence for the risk-response outcomes of the NIST AI RMF (MANAGE 1.3) and the safety expectations of MEASURE 2.6.

The Deterministic Boundaries

The Consequence Router does not make suggestions. It operates as a physical circuit breaker between the agent's intelligence layer and its capacity to act. Every evaluation terminates in one of three hard mathematical boundaries: PASS, FLAG, or HALT.

1. The PASS Boundary (Machine-Speed Execution) When an intelligence signal cleanly satisfies all eight streams—demonstrating high calibrated source reliability, robust independent corroboration, temporal relevance, and an unbroken cryptographic state—the Consequence Router renders a PASS verdict.

The Doubt Index remains negligible. The execution boundary opens, and the agentic orchestrator is authorized to act at machine speed. The enterprise can trust this autonomous transition because it is backed by mathematical friction, not probabilistic hope.

2. The FLAG Boundary (Mandatory Human Escalation) If the engine detects moderate anomalies—such as a spike in the Doubt Index due to inter-stream contradiction, or intelligence degrading past its temporal half-life—it routes the signal to the FLAG boundary.

Execution is immediately suspended. By escalating the decision to a human reviewer, the system directly supports the **intervention and deactivation protocols of the NIST AI RMF (MANAGE 2.4)**. Crucially, the engine does not just hand the reviewer a warning; it hands them the exact, per-stream breakdown of *why* the intelligence was contested, providing a structured, mathematically sound basis for human judgment rather than asking them to interpret a raw probability.

3. The HALT Boundary (The Kill Switch) The HALT boundary is the ultimate enforcement of the Multiplicative Collapse rule. It is triggered by foundational failures: an adversarial trajectory detection (Stream 5), a Curveball corroboration attack (Stream 6), or a cryptographic hash mismatch in the registry (Stream 8).

When a HALT is triggered, the execution pathway is severed. The engine collapses the evidentiary weight to zero. The agent is entirely locked out of the state transition. There is no override, no prompt-injected bypass, and no probabilistic leniency.

Causal Attribution: The Shapley Decomposition

When an agentic system executes a high-consequence action, or when a HALT triggers a massive operational alert, an auditor will inevitably ask: *What exact piece of intelligence caused this system to act?*

If you are using a neural network or a black-box LLM, you cannot answer that question. You can only guess.

The Verdict Weight Consequence Router solves this through **Exact Shapley Value Decomposition**. Borrowed from cooperative game theory, Shapley values provide a mathematically principled method for distributing the "credit" or "blame" for an outcome across a coalition of players. In our engine, the "players" are the eight streams.

For every verdict rendered, the engine computes the exact marginal contribution of each stream to the final execution decision. This decomposition strictly satisfies four mathematical axioms (Efficiency, Symmetry, Dummy, and Additivity), meaning the attribution is exact, not an approximation.

- If a transaction was blocked, the Shapley output will prove mathematically that Stream 2 (Lack of Corroboration) contributed 68% to the HALT verdict, while Stream 5 (Adversarial Trajectory) contributed 32%.
- If an action was authorized, the output will prove exactly how much Stream 1 (Source

Reliability) and Stream 4 (Historical Accuracy) drove the PASS verdict.

This is the ultimate audit receipt. It provides absolute causal transparency, perfectly aligning with **NIST MEASURE 2.9 (Explainability)**. You are no longer defending a black-box decision; you are presenting a mathematically verifiable record of execution.

Chapter 7

The Proof

Everything in this book so far has been architecture. We have described what the engine does, how the streams compose, and why Multiplicative Collapse is the only honest standard for execution control. None of that matters if the math does not survive contact with real data.

This chapter presents that contact. We take the VERDICT WEIGHT™ engine, deploy it against real-world clinical AI predictions on real patient chest X-rays, and measure exactly what happens. No synthetic data. No cherry-picked scenarios. Real models, real images, real pathologies.

The results are not perfect. They are honest. And they reveal something about the framework that matters more than any single metric: VERDICT WEIGHT™ improves calibration *without access to any ground-truth labels*. It does not learn a correction curve from answers it already knows. It evaluates the evidence pipeline itself—and the evidence pipeline, it turns out, carries enough information to transform an unreliable AI system into a dramatically better-calibrated one.

The Experiment: CheXpert Radiology Validation

CheXpert is one of the most widely used benchmark datasets in medical AI. It contains hundreds of thousands of chest X-rays from Stanford Hospital, labeled across fourteen pathologies by expert radiologists. It is the standard proving ground for clinical AI systems.

We trained three independent deep learning models—DenseNet-121, EfficientNet-B0, and ResNet-50—on a stratified 50,000-image training sample from CheXpert. We then evaluated them on a held-out 202-image validation set across five competition pathologies: Atelectasis, Cardiomegaly, Consolidation, Edema, and Pleural Effusion.

Here is what we found before VERDICT WEIGHT™ was applied: the raw models were badly calibrated. The macro-averaged Expected Calibration Error was 0.305—meaning the stated confidence of these models bore almost no relationship to their actual accuracy. A model claiming 80% confidence was wrong far more often than 20% of the time. If you deployed these models to advise a radiologist, the numbers on the screen would be actively misleading.

This is not a failure of these specific models. This is the baseline state of clinical AI. Most published radiology AI systems exhibit similar calibration deficits. The industry has accepted this because the standard response—post-hoc recalibration using

Platt scaling or isotonic regression—can fix the numbers. But those methods require labeled validation data. They require you to already know the right answers.

VERDICT WEIGHT™ does not know the right answers. It was never given the labels. It evaluated the *evidence pipeline*—the reliability of each model, the agreement between them, their temporal freshness, and their historical track records—and produced a Consequence Weight for every prediction.

The Headline: Label-Free Calibration

Applied to the uncalibrated model outputs, the VERDICT WEIGHT™ Government deployment tier reduced the Expected Calibration Error from 0.305 to 0.089—a 70.8% reduction—without access to a single ground-truth label. The Brier score dropped from 0.170 to 0.063, a 63% reduction, and reliability error fell from 0.117 to 0.009, a 92.6% reduction. The AUROC is preserved at 0.812: VERDICT WEIGHT™ improves how honest the system's confidence is without claiming to improve its discrimination.

To understand what these numbers mean, consider the standard post-hoc calibration baseline. Platt scaling, which *does* have access to the labels, achieves an ECE of 0.007 and a Brier score of 0.056.

VERDICT WEIGHT™ reaches 0.089 ECE and 0.063 Brier without labels. It closes over 70% of the calibration gap using evidence alone.

Calibration Progression

Method	ECE	Brier	REL	Labels?
Uncalibrated (Simple Avg)	0.305	0.170	0.117	No
VW GOV (Raw Inputs)	0.089	0.063	0.009	No
Platt Scaling (5-fold CV)	0.007	0.056	0.001	Yes
VW + Platt (Combined)	0.009	0.056	0.001	Yes

The critical column is REL—reliability error. The uncalibrated baseline has a reliability error of 0.117. VERDICT WEIGHT™ reduces it to 0.009. That is a 92.6% reduction in the mathematical measure of how much the stated confidence deviates from reality. The framework is not guessing better—it is fundamentally improving how honest the system's confidence is.

Composability: Not a Replacement, a Layer

The most important finding is not that VERDICT WEIGHT™ matches Platt scaling. It is what happens when you combine them.

Post-hoc calibration and evidence-based confidence weighting are not competing approaches. They are complementary layers in a calibration stack. Platt scaling corrects the probability scale. VERDICT WEIGHT™ provides the evidence architecture that explains *why* a prediction should be trusted.

When we Platt-scale each model's predictions via 5-fold cross-validation, compute the calibrated ensemble average, and then apply VERDICT WEIGHT™ weighting, the combined pipeline achieves an ECE of 0.009 and a Brier of 0.056—matching standalone Platt—while adding something Platt cannot provide: decomposable confidence. Every prediction carries four stream scores, a Doubt Index, and a Consequence Weight. A radiologist or auditor can inspect exactly *why* a prediction received high or low confidence. Platt gives you a better number. VERDICT WEIGHT™ gives you a better number you can explain.

These results are stable. We tested across five random cross-validation seeds, three fold configurations, and all five pathologies individually. The ECE standard deviation is 0.003 across seeds. The framework does not depend on a lucky split.

The Hidden Danger: Correlated Failure

Here is the finding that justifies Stream 6 (Source Independence) more than any architectural argument ever could.

All three of our models were trained on the same 50,000-image sample from CheXpert, evaluated on the same 202-image validation set, and initialized from ImageNet pretrained weights. They are different architectures—DenseNet, EfficientNet, ResNet—but they share the same data lineage.

We computed pairwise Pearson correlation of squared prediction errors across all five pathologies. The result: a grand mean correlation of **r = 0.820**. A permutation test with 10,000 shuffles produced a z-score of 44.5 ($p < 0.0001$). The models' errors are not independent. They are massively, structurally correlated.

This means that when all three models agree—which an ensemble framework interprets as high confidence—a substantial fraction of those agreements are *shared errors*. All three models are confidently wrong on the exact same images. Standard ensemble averaging amplifies this shared overconfidence. It treats correlated agreement as independent corroboration.

This is the Curveball problem in a clinical setting. Not a hostile intelligence agency manufacturing fake sources—but a medical AI deployment where every "independent" model was trained on the same data. The corroboration is an echo, not a confirmation.

Pairwise Error Correlation

Model Pair	Atel.	Card.	Cons.	Edema	Pl.Eff.	Macro
DenseNet – EfficientNet	0.730	0.782	0.856	0.847	0.799	0.803
DenseNet – ResNet	0.736	0.840	0.815	0.879	0.852	0.824
EfficientNet – ResNet	0.855	0.826	0.757	0.853	0.867	0.832

Grand mean: $r = 0.820$. Permutation test $z = 44.5$, $p < 0.0001$. The correlation is 44.5 standard deviations above the null distribution. Shared failure modes are pervasive and statistically unambiguous.

In a realistic clinical deployment, radiology AI systems from different vendors would differ in training data, architecture, regulatory clearance pathway, and institutional validation history. The correlation we measured—$r = 0.820$—represents an upper bound. As source diversity increases, the Doubt Index gains discriminative power and the framework's value over simple ensemble averaging grows. But the finding stands: if you do not verify source independence, your corroboration is a lie.

Per-Pathology: Where the Evidence Hits

The calibration gain is not an artifact of pooling—it holds across every competition pathology individually. The table below reports Expected Calibration Error for each of the five, uncalibrated ensemble versus VERDICT WEIGHT™ GOV. Every pathology improves substantially, and the gains are largest where the raw ensemble was worst calibrated.

Pathology	Uncalibrated ECE	VW GOV ECE
Atelectasis	0.419	0.106
Cardiomegaly	0.244	0.062
Consolidation	0.336	0.093
Edema	0.287	0.088
Pleural Effusion	0.239	0.096
Macro (5-pathology)	**0.305**	**0.089**

The pattern is consistent across findings: every competition pathology sees its Expected Calibration Error fall by roughly 60 to 75%, from a macro-averaged 0.305 to 0.089. Atelectasis—the worst-calibrated finding in the raw ensemble at 0.419—improves to 0.106, and Pleural Effusion, already the best-calibrated at 0.239, improves to 0.096. The gain is broad-based, not driven by any single pathology, and it is achieved without labels.

The Honest Caveat

The numbers above are real, reproducible, and independently verifiable. Every result traces to a single Redivis table (val_predictions_50k_seed42_v12:f9zf). Any researcher can reproduce them.

But the results require context that this book will not hide from you.

The Government deployment tier's improvement is driven in part by probability compression. The GOV configuration caps the Consequence Weight at approximately 0.289. This means the framework never expresses high confidence in a positive finding —which makes the Brier score look excellent, because you are never confidently wrong. But it also means you are never confidently right at clinical decision thresholds.

The AUROC is preserved at 0.812, which proves the *ranking* is intact. The framework knows what is more suspicious than what. But the absolute confidence values in GOV configuration cannot drive clinical accept/reject decisions at standard thresholds. The framework's real clinical value in this configuration is as a **confidence-bounding layer**—a mechanism that tells you how much to trust the evidence, not whether the patient has the disease.

The validation set comprises 202 images. With 10 calibration bins, mean bin occupancy is approximately 20 per pathology. Bootstrap 95% confidence intervals quantify estimation uncertainty throughout, but external validation on MIMIC-CXR or PadChest is required to confirm generalization.

These limitations are not buried in an appendix. They are here, in the chapter about proof, because this is what intellectual honesty looks like in execution control. If you cannot state the boundaries of your evidence with the same precision you state the evidence itself, you are doing compliance theater—the exact thing this methodology was built to eliminate.

What This Proves

The CheXpert validation establishes four things:

First, VERDICT WEIGHT™ improves calibration without access to ground-truth labels. This is the fundamental value proposition. In any deployment where labeled validation data is scarce, expensive, or nonexistent—which describes most real-world AI deployments—the framework provides calibration improvement from the evidence pipeline alone.

Second, the framework composes with existing calibration methods rather than competing with them. VW + Platt matches Platt while adding decomposable confidence, lineage awareness, and an audit trail. You do not have to choose between statistical calibration and evidence-based governance. You get both.

Third, correlated failure modes are real, measurable, and dangerous. The r = 0.820 error correlation proves that source independence is not a theoretical concern—it is an empirical fact. Any framework that treats model agreement as independent corroboration without verifying lineage is structurally vulnerable.

Fourth, the framework's limitations are bounded and documented. The probability compression effect, the GOV configuration ceiling, the 202-image sample size—these are stated, quantified, and traceable. The proof is not that the framework is perfect. The proof is that the framework is honest, and that honest evidence-based weighting outperforms uncalibrated probability in every measurable dimension.

The architecture described in Chapters 1 through 5 is not a theoretical exercise. It is a working system, validated on real clinical data. The question is no longer whether deterministic execution control works. The question is how long your enterprise will continue operating without it.

Data Provenance

All results computed from Redivis table:
val_predictions_50k_seed42_v12:f9zf

Models: DenseNet-121, EfficientNet-B0, ResNet-50 (50K training sample, seed 42/43/44)

VW weighting: verdict-weight v1.2.0 (PyPI), HEALTHCARE_DIAGNOSTIC context

Validation: 5 seeds × 3 fold counts × 5 pathologies × 2,000 bootstrap resamples

Published: SSRN (DOI: 10.2139/ssrn.6532658) | Zenodo (DOI: 10.5281/zenodo.19447547)

Chapter 8
The Cybersecurity Proof

Chapter 7 proved the framework on clinical radiology data. The natural objection is that medical imaging is a narrow domain—perhaps the calibration improvements depend on the specific structure of chest X-ray predictions and would not generalize to other high-consequence environments.

This chapter answers that objection by deploying VERDICT WEIGHT™ against real-world cybersecurity vulnerability data: 6,404 CVEs drawn from the NIST National Vulnerability Database, scored against the CISA Known Exploited Vulnerabilities catalog as ground truth. The domain is entirely different. The data sources are entirely different. The question is the same: does the framework improve the trustworthiness of autonomous assessments?

It does. But the path to proving it taught us something important about what the framework actually detects, and it is not what we initially expected.

Section 1: The Signal Selection Lesson

Our first validation attempt used three data sources as "models"—direct analogs to the three neural networks in the CheXpert experiment:

EPSS (Exploit Prediction Scoring System)—a probabilistic machine learning model that predicts the likelihood a vulnerability will be exploited within 30 days. This is a true probability, directly analogous to a DenseNet outputting a pathology prediction.

CVSS Base Score (Common Vulnerability Scoring System)—the industry-standard severity rating assigned by NIST analysts, normalized from 0–10 to 0–1.

CVSS Exploitability Sub-Score—a component of the CVSS score capturing how easy the vulnerability is to exploit.

The ground truth was CISA KEV membership: was this vulnerability actually exploited in the wild?

The results were not what we expected.

First Run: Calibration Metrics

Method	ECE	Brier	AUROC
EPSS (Raw)	0.110	0.086	0.993
CVSS Base (Normalized)	0.401	0.402	0.369
Simple Average	0.285	0.171	0.829
VW GOV	0.385	0.251	0.829

EPSS alone was a near-perfect predictor (AUROC 0.993). CVSS, the most widely used vulnerability scoring system in the world, showed an apparent AUROC of 0.369—seemingly anti-predictive. VERDICT WEIGHT™ performed *worse* than raw EPSS because it was dutifully incorporating two degraded signals into the composite score.

Our first instinct was to declare CVSS anti-predictive. That would have been a compelling headline. It would also have been wrong.

When we investigated, we found that only 33% of KEV CVEs had real CVSS data from the NVD pull—the remainder carried default fill values of 0.5. The anti-correlation was a coverage artifact. For the 5,067 CVEs with real CVSS scores, the AUROC was 0.756—reasonable, not inverted.

The framework did not fail. It told us the truth: garbage in, garbage out. No weighting scheme, no matter how sophisticated, can rescue fundamentally degraded input signals. VERDICT WEIGHT™ governs the *sufficiency* of evidence. It does not manufacture evidence that does not exist.

What the Doubt Index Actually Detected

But the first run was not a waste. It produced a finding that proved more valuable than a clean calibration win.

When we examined the Doubt Index distribution, we found a near-perfect separation:

DI Range	CVE Count	KEV Rate
DI < 0.20	908	99.7%
DI >= 0.20	5,496	12.7%

Of those 908 CVEs, 99.7% were genuine known-exploited vulnerabilities—near-perfect precision. The band did not contain every exploited CVE; it held 905 of the 1,601, about 57% recall. This is a high-purity triage filter, not a catch-everything one: an analyst could focus on those 908—an 86% reduction in volume—and find almost all of them genuine, while the remaining exploited CVEs stay distributed across the larger, noisier pool. The framework isolated the assessments the evidence pipeline most agreed on, achieved without access to the KEV labels.

The mechanism was not inter-source contradiction. The correlation between DI and actual source disagreement was only r = 0.11. What DI measured was *evidence consensus*. When EPSS was high and the other signals were present and aligned, DI dropped below 0.20. When the evidence pipeline was sparse—when CVSS data was missing for 67% of KEV entries—DI rose, and the framework correctly flagged those assessments as structurally uncertain rather than falsely confident.

The framework did not tell us which vulnerabilities were dangerous. It told us which vulnerability assessments to trust. That distinction matters.

Signal Recovery Under Noise

On the 805 highest-stakes CVEs (EPSS > 0.7, 100% KEV rate), VERDICT WEIGHT™ GOV scored 0.7446 against simple averaging at 0.6649—a recovery of +0.0797. The framework partially restored the EPSS signal that CVSS noise was drowning out. It did not match raw EPSS (0.9005) because it was correctly penalizing for weak corroboration. That penalty is the feature, not the bug.

But the first run made clear that signal selection matters more than signal weighting. No framework should be evaluated on inputs chosen to fail. So we rebuilt the experiment.

Section 2: Three Independent Sources

For the second validation, we replaced CVSS with two sources that are genuinely predictive of real-world exploitation and genuinely independent of EPSS:

Source 1: EPSS—the same probabilistic ML model, retained as the primary signal. Sourced from FIRST.org.

Source 2: ExploitDB—a curated database maintained by Offensive Security containing verified proof-of-concept exploits. The signal: does a public, working exploit exist for this CVE? Binary, high-quality, maintained by a different organization using a different methodology.

Source 3: GitHub PoC Repositories—a community-maintained mapping of CVEs to proof-of-concept code published on GitHub. The signal: how many independent researchers have published exploit code? This measures exploit maturity through organic community behavior.

Three signals. Three different methodologies. Three different organizations. Genuine independence.

Calibration Results

Method	ECE	Brier	AUROC
EPSS (Raw)	0.111	0.087	0.993
ExploitDB	0.182	0.182	0.636
GitHub PoC	0.216	0.198	0.728
Simple Average	0.169	0.134	0.991
VW GOV	0.135	0.099	0.991
VW APT	0.132	0.098	0.991

VERDICT WEIGHT™ GOV achieves a Brier score of 0.099 against simple averaging at 0.134—a **26.2% improvement**. The APT deployment profile (tuned for advanced persistent threat scenarios) achieves 26.5%. Both preserve the AUROC at 0.991 while substantially improving calibration. That near-perfect AUROC is partly a property of this benchmark—KEV inclusion correlates with the exploitation signals by construction, so any method that aggregates these channels will rank them well. The load-bearing result here is the calibration gain, not the discrimination number.

This is the parallel to CheXpert. In radiology, the framework improved Brier by 63% label-free. In cybersecurity, it improves Brier by 26% with genuinely corroborating sources. The improvement is smaller because EPSS is already an excellent predictor—and in fact EPSS alone posts a lower Brier (0.087) than the fused VERDICT WEIGHT™ output (0.099). The framework does not beat the best single source here; it beats naive averaging of unequal sources, and it does so while keeping the fused confidence decomposable and auditable. When one source is known to dominate, use it. The framework earns its place when you must combine heterogeneous sources of unknown relative quality and need calibrated, explainable confidence without labels. But the direction is the same: evidence-based confidence weighting produces better-calibrated outputs than naive averaging, and it does so by evaluating the evidence pipeline rather than learning a correction curve.

The Doubt Index: Working as Designed

With three genuinely independent sources, the Doubt Index exhibited exactly the behavior the architecture predicts:

DI Range	CVE Count	KEV Rate
DI [0.1, 0.5)	1,112	99.0–100%
DI [0.5, 1.0)	5,292	9.4%

Clean separation at DI = 0.5. Below that threshold, virtually every CVE is a known exploited vulnerability. Above it, fewer than one in ten.

But the critical finding is not the separation itself—it is what drives it. We examined three signal profiles:

Signal Profile	N	DI	CW	KEV Rate
All 3 agree high	266	0.150	0.736	100%
EPSS high, no public exploit	231	0.346	0.451	100%
All 3 agree low	4,988	0.606	0.014	5.7%

When all three sources converge—EPSS predicts high exploitability, a public exploit exists in ExploitDB, and GitHub PoC repositories confirm exploit maturity—the Doubt Index drops to 0.150 and the Consequence Weight rises to 0.736. The evidence pipeline agrees. The framework is confident.

When EPSS flags a vulnerability as high-risk but no public exploit exists and no PoC repositories confirm it, the Doubt Index rises 131% to 0.346 and the

Consequence Weight is suppressed 39% to 0.451. The signal is real but unconfirmed. The framework says: *we believe this, but we are not certain.*

When all three sources agree nothing is happening—low EPSS, no exploit, no PoC—the Doubt Index rises to 0.606 and the Consequence Weight collapses to 0.014. The evidence pipeline has nothing to report. The framework has nothing to trust.

> *The framework did not change the underlying probability—it changed its confidence in that probability.*

That is the entire value proposition in one sentence. The framework does not predict. It evaluates the evidence behind the prediction and tells you how much to trust it.

What the Two Sections Prove Together

First, signal selection is a prerequisite for signal weighting. The CVSS run proved that no framework can rescue anti-predictive or incomplete inputs. VERDICT WEIGHT™ will faithfully evaluate whatever evidence you give it—including evidence that is structurally degraded. The framework is honest, not magical. Section 1 is the proof of that honesty.

Second, when given genuinely independent, genuinely predictive sources, the framework improves calibration by 26% over naive averaging while preserving the AUROC at 0.991 (near-perfect here largely because this benchmark's labels correlate with its inputs by construction). The improvement is driven by the same mechanism as in CheXpert: evidence-based weighting of source reliability and corroboration, not learned correction curves.

Third, the Doubt Index is a structurally valid triage mechanism across domains. In radiology, low DI identified the predictions most likely to be correct. In cybersecurity, low DI identified the vulnerabilities most likely to be exploited. The mechanism is domain-agnostic: when independent evidence streams converge, the assessment is trustworthy. When they diverge or are absent, it is not.

Fourth, the framework crosses domains. Radiology and cybersecurity share nothing—different data types, different source structures, different failure modes, different regulatory frameworks. The architecture works in both because it operates at the evidence layer, not the domain layer. It does not understand chest X-rays or CVE scores. It understands source reliability, corroboration, and temporal freshness. Those properties are universal.

The Cross-Domain Standard

Between Chapter 7 and this chapter, VERDICT WEIGHT™ has been validated against 2,828 clinical radiology predictions (202 images across all 14 CheXpert pathologies, with headline calibration reported on the five competition pathologies) and 6,404 real-world cybersecurity vulnerability assessments. In both domains, the same 8-stream architecture, the same Consequence Router, and the same Doubt Index produced measurably better-calibrated outputs than uncalibrated baselines and naive averaging.

The framework does not know what a lung opacity is. It does not know what a buffer overflow is. It knows whether the evidence behind those assessments is reliable, corroborated, and temporally fresh. And it turns out that is enough.

Data Provenance

Section 1: NIST NVD + CISA KEV + FIRST EPSS | 6,404 CVEs (1,601 KEV positive) | Registry v1.2.0 (cybersecurity)

Section 2: FIRST EPSS + ExploitDB (Offensive Security) + GitHub PoC (nomi-sec) | Same 6,404 CVE cohort | Registry v1.2.0

Ground truth: CISA Known Exploited Vulnerabilities catalog

VW weighting: verdict-weight v1.2.0 | CYBERSECURITY context | Source Registry v1.2.0 (PRIVATE)

Bootstrap: 2,000 resamples, seed 42 | All results independently reproducible from source data

Conclusion
The Evidence Standard

The era of discretionary AI governance is over.

For the last three years, the enterprise AI industry has operated under the delusion that if a foundation model is aligned, its actions are safe. We have treated security as a polite suggestion, asking probabilistic models to police their own hallucinations and relying on heuristic guardrails to catch the fallout.

As global regulatory frameworks transition from theoretical guidelines to enforceable law, that paradigm has become a massive structural liability. The NIST AI RMF and emerging regulatory frameworks do not penalize bad thoughts; they penalize unauthorized, catastrophic execution. If your enterprise cannot mathematically prove exactly why an autonomous agent was allowed to alter a state, transfer a fund, or trigger a system alert, you do not have an execution boundary. You have an open door for algorithmic contagion.

The Verdict Weight methodology was engineered to slam that door shut.

The Evidence Standard Summarized

We have shifted the perimeter from *generation* to *evidence*. By deploying the 8-stream execution control engine, we replace the probabilistic guessing game with a deterministic, mathematical standard.

1. **Tier 1 (The Commercial Baseline):** We ruthlessly interrogate the data. By decoupling calibrated source reliability (Stream 1) from independent corroboration (Stream 2), enforcing temporal decay (Stream 3), tracking historical Bayesian accuracy (Stream 4), and punishing contradictions via the Doubt Index, we ensure that the intelligence is structurally sound before it is evaluated.
2. **Tier 2 (The Adversarial Moat):** We neutralize the trajectory. Stream 5 calculates cross-temporal consistency, detecting the spike-then-collapse fingerprints of adversarial data poisoning and routing the contagion to human review before it can bypass the logic gates.
3. **Tier 3 (Cryptographic Government Assurance):** We lock the infrastructure. Stream 6 verifies physical independence, Stream 7 mints the immutable, SHA-256 execution receipt, and Stream 8 enforces the ultimate fail-safe—collapsing the engine to zero if the underlying registry is compromised.

The culmination is the **Consequence Router**. The engine renders a deterministic verdict—PASS, FLAG, or HALT—backed by an exact Shapley decomposition of the causal attribution. There are no percentages. There are no vibes. There is only mathematical friction and verifiable proof.

Your Marching Orders: Enterprise Integration

Understanding the math is only the first step. You must now operationalize the Evidence Standard across your enterprise AI architecture. If you are a Risk Officer, a CISO, or an AI Engineering Lead, these are your immediate marching orders:

1. Audit Your Current Boundary Assume your current guardrails will fail a regulatory audit. Pull the Verdict Weight Evidence Checklist for the NIST AI RMF. Force your engineering teams to answer the binary questions. If they cannot prove cryptographic provenance or if they rely on an LLM to evaluate prompt injections, flag the architecture as non-compliant immediately.

2. Deploy the Engine Upstream Do not attempt to bolt this logic onto the end of a generative output. The Verdict Weight engine must sit upstream of the agentic orchestrator. Whether you are using the Model Context Protocol (MCP) or a custom RAG pipeline, the engine must intercept the intelligence payload, run the 8-stream interrogation, and render the verdict *before* the context window is populated.

3. Enforce the Cryptographic Ledger Stop logging prompts and start hashing states. Configure your pipeline to capture the Stream 7 Cryptographic Provenance State for every single execution evaluation. When the regulator, the auditor, or the board of directors demands an explanation for an autonomous action, you will not hand them a log file; you will hand them the immutable Shapley decomposition receipt.

4. Accept the Multiplicative Collapse You must train your organization to accept that the HALT boundary is non-negotiable. When the engine detects a registry hash mismatch or an adversarial trajectory, it will sever the execution pathway. Do not engineer bypass routes for executives. Do not create discretionary overrides. If you compromise the deterministic nature of the fail-safe, you absorb the liability of the contagion.

The Final Verdict

Agentic AI is the most powerful operational leverage ever introduced to the enterprise. But leverage without control is just a faster way to break your own infrastructure.

Stop asking your AI to guess if it is safe. Force it to prove that the evidence holds. Implement the engine, enforce the boundaries, and let the math govern the execution.

www.ingramcontent.com/pod-product-compliance
Lightning Source LLC
LaVergne TN
LVHW080531210926
848481LV00021B/2553